How We Use

Wood

Chris Oxlade

Raintree

Chicago, Illinois

© 2004 Raintree
Published by Raintree, a division of Reed Elsevier, Inc.
Chicago, IL 60602
Customer Service 888-363-4266
Visit our website at www.raintreelibrary.com

For more information address the publisher:
Raintree, 100 N. LaSalle, Suite 1200, Chicago IL 60602

Printed and bound in China by the South China
   Printing Company

08 07 06 05 04
10 9 8 7 6 5 4 3 2 1

**Library of Congress Cataloging-in-Publication
Data:**
Oxlade, Chris.
  How we use wood / Chris Oxlade.
     p. cm. -- (Using materials)
Includes bibliographical references and index.
Contents: Wood and its properties -- Where does wood
come from? -- Wood for fuel -- Wood for building --
Wooden furniture -- Wood for decoration -- Weaving
wood -- Wood for boards -- Wood for paper -- Wood
afloat -- Preserving wood -- Wood for objects -- Wood
and the environment.
  ISBN 1-4109-0598-5 (hc) 1-4109-0997-2 (pb)
 1. Wood--Juvenile literature. [1. Wood.] I. Title. II.
Series:
Oxlade, Chris. Using materials.
  TA419.O94 2004
  620.1'2--dc21
                              2003007432

COVER

Acknowledgments
The publishers would like to thank the following
for permission to reproduce photographs:
p. 4 Dr. Richard Kessel & Dr. C. Shih/Visuals
Unlimited; p. 5 Owaki-Kulla/Corbis; p. 6 Inga
Spence/Visuals Unlimited; p.7 Charles O'Rear/Corbis;
p. 8 Peter Kubal; p. 9 Rick Wetherbeep; p. 10 Harcourt;
p.11 Zandria Muench Beraldo; p.12 David
Wrobel/Visuals Unlimited; p.13 Helene Rogers/Art
Directors & TRIP; p.14 M. Angelo/Corbis; p.15 Michael
Boys/Corbis; p. 16 Jonah Calinawan; p. 17 Mark
Gibson; p.18 Dwight Kuhn; p.19 Michele
Burgess/Visuals Unlimited; p.20 Dave G.
Houser/Corbis; p.21 Hans Georg Roth/Corbis; p.22 Joe
McDonald/Visuals Unlimited; p.24 Robert
Estall/Corbis; p.25 Yves Tzaud; pp.26, 29 Corbis; p.27
Jeff Greenberg/Visuals Unlimited; p.28 Greg
Williams/Heinemann Library.

Cover photographs reproduced with permission
of Corbis.

Every effort has been made to contact copyright
holders of any material reproduced in this book. Any
omissions will be rectified in subsequent printings if
notice is given to the publishers.

Contents

Any words appearing in bold, **like this,** are explained
in the Glossary.

Wood and Its Properties

All the things we use at home, school, and work are made from materials. Wood is a material. We use wood for all sorts of different jobs. You can see wood being used wherever you go. For example, in the kitchen you might see wooden spoons. On a building site you could see whole houses being made from wood.

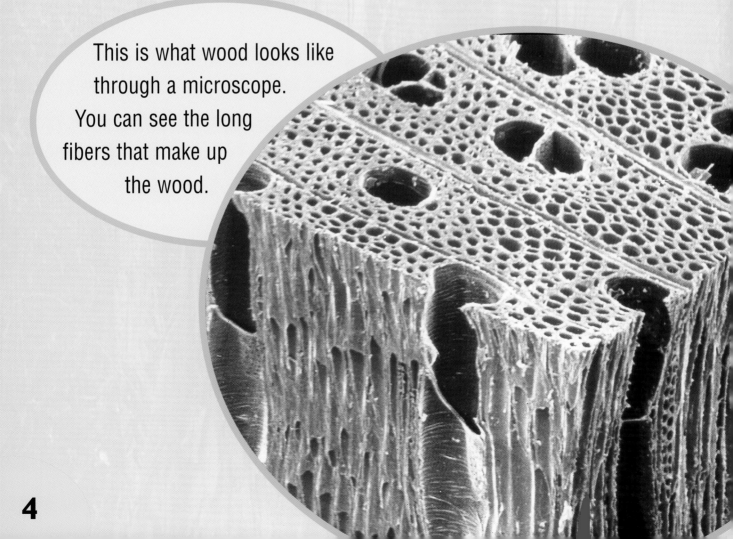

This is what wood looks like through a microscope. You can see the long fibers that make up the wood.

Here, wood has been used to make chairs, walls, and floors.

Properties tell us what a material is like. Most wood is strong. Some wood is light and some is heavy. Some wood is soft and some is hard. Wood comes in different colors and with different patterns. All wood is made up of long **fibers.** These make up the lines in wood.

Don't use it!

*The different properties of materials make them useful for different jobs. For example, wood **rots** if it gets damp. So we would not make an underground pipe from wood. It would soon rot and leak.*

Where Does Wood Come From?

Wood is a **natural** material. It comes from trees. The trunk and branches of a tree are made of wood. The leaves are not. There are thousands of different kinds of trees. Each kind of tree gives us wood with different **properties.** For example, the wood from a pine tree is light in color and very soft, but the wood from a walnut tree is dark and hard.

This tree was grown so people can use its wood. A new tree will be planted in its place.

The forestry industry

Huge areas of land are used to grow trees for wood. This is called the

Wood in history

*Wood was probably the first material that people learned how to use. For thousands of years people have made wooden tools, built homes and ships from wood, and burned wood as a **fuel.***

forestry **industry.** Trees take between 20 and 50 years to grow big enough to be used for wood. Then they are cut down and the branches are cut off. The trunks are carried to a sawmill. Here they are cut into long pieces and sold.

Planks cut from trees have to dry out before they are used.

Wood for Fuel

A **fuel** is a material that we burn to make heat for cooking and keeping warm. Wood is a good fuel which means that it burns well. Wood only catches fire quickly when it is dry. Wet wood is hard to light. Logs for fires are made by cutting up tree trunks and thick branches. They need to be dried so they will burn well.

We use thin strips of wood called kindling to get a fire started because they catch fire quickly.

This stove is designed to burn wood instead of other fuels.

Finding wood

In remote places fuels like gas, oil, and coal are difficult to get or too expensive to buy. Here people use wood for cooking, heating, and for making light. Sometimes their homes are surrounded by forests, so wood is easy to find. But sometimes all the wood has been used up and people have to walk many miles searching for twigs and branches.

Charcoal

If wood is burned with only a little air it smolders slowly. Instead of turning to ash it turns into a black solid called charcoal. Artists use charcoal for drawing. Charcoal is a good fuel for barbecues because it burns slowly.

Wood for Building

Wood is a strong material. Thin pieces of wood bend easily but thick pieces of wood are hard to bend. We use long, thicker pieces of wood to make frames inside buildings. For example, wooden frames called trusses hold up roofs in houses. In some houses the walls are held up by a wooden frame, too.

These wooden frames provide the support of a house.

In some places in the world wood is used to build homes from tree trunks and branches.

Builders also use strong pieces of wood to hold up floors and for floorboards. Some houses have walls covered in wood called siding. It keeps the wind and rain out of the house. Wooden siding also looks good because wood is a **natural** material.

Don't use it!
Wood is strong enough for building houses, but not for building very large structures, such as skyscrapers and long bridges. Materials like concrete and steel are much stronger.

Wooden Furniture

The **properties** of wood mean that it is good for making furniture. Wood is sturdy and easy to cut. People also like its **natural** appearance. We make chairs, tables, beds, cupboards, and shelves from wood. Soft furniture, such as sofas, have wooden frames hidden inside. We also use wood for doors.

All the parts of this dresser are made from wood.

The edges of these two pieces of wood are held together by a dovetail joint.

Furniture is made from many different kinds of wood. For example, pine makes light, cheap furniture. Mahogany makes beautiful dark furniture, but it is expensive because it is rare and takes a long time to grow.

Don't use it!

Wood must be dried properly before it is made into furniture. Furniture made from damp wood can twist and bend as the wood dries out quickly in a warm house. This is called warping.

Working with wood

Wood is easy to cut and shape with metal tools such as saws and chisels. Pieces are joined together with screws, nails, bolts, or glue. The wooden parts of most pieces of furniture are cut out and shaped by machines.

Wood for Decoration

Wood comes in different colors. Some woods are almost white. Some are yellow, but most are brown. A few sorts are red. When a tree trunk or a branch is cut lengthwise you see an interesting pattern of long wavy stripes. This pattern is called **grain.** Dark circles in the grain are called knots. The color and pattern in the grain of wood make it good for decoration.

Burlwood is a type of wood with beautiful swirls.

The patterns on this chair are made from skillfully cut pieces of veneer.

Dirt and water spoil wood. We can protect wood with **chemicals** such as oil and wax. They also help to make the grain show up clearly. Painting wood with **varnish** helps to stop it from getting scratched by harder materials such as metal or glass.

Veneers

A **veneer** is a very thin sheet of wood. We use veneers with beautiful color and grain to cover plain wood or board. Craftspeople make patterns and pictures on furniture using small pieces of veneer made from different woods.

Wood for Boards

A board is a wide, flat, thin piece of wood. We can cut only narrow boards, such as floorboards, from tree trunks because the trunks are narrow. Floorboards are strong and less expensive than metal or concrete.

Other types of wooden board are made by glueing smaller bits of wood together. Plywood is made by glueing thin strips of wood together in layers. The wood is laid so that the **grain** in one layer is at right angles to the grain in the layers above and below. This makes plywood very strong. We use plywood to make doors and floors.

This is the edge of a piece of plywood. You can see the edge of some layers and the ends of others.

This chipboard is being used as a frame for a house.

Boards from bits

Chipboard and fiberboard come from waste wood from a sawmill. Chipboard is made from small pieces of wood called chips. They are mixed with glue and the mixture is pressed flat until the glue dries. We use chipboard for making cupboards and worktops. Chipboard can be covered with **veneer** to make it look like real wood.

Wood for Paper

The paper in this book is made from wood. So is newspaper, writing paper, paper for envelopes, and cardboard. Paper is made from wood **fibers,** which are pressed together into a very thin sheet. Wood is used to make paper because the fibers are cheap and plentiful.

This is an edge of ripped paper seen through a microscope. You can see the ends of the wood fibers.

Wood pulp
has been put onto a wire
mesh to make this paper.

Wood to paper

One out of every ten trees that are cut down is used to make paper. The wood is broken into tiny pieces and mixed with **chemicals** that separate the fibers from each other. This makes a thick fiber paste called wood pulp. The pulp is spread out, rolled flat, and then dried to make the finished paper.

Wood for Boats

Wood floats on water. This is because a piece of wood weighs less than the same amount of water. Very light wood, such as balsa wood, only sinks into the water a tiny bit. Heavy wood, such as teak, sinks in much further before it floats. The simplest boats are rafts made by tying logs or planks together.

This raft in Fiji is made from long, thin logs.

This man is shaping part of a traditional wooden boat.

Boat building

Small boats are made from wood because wood is easy to cut, shape, and join with simple tools. The pieces of wood are joined together tightly to stop water from leaking through the gaps between them. If the wood does get damp, it gets bigger and fills the gaps. Wooden boats must be painted or varnished to make them **waterproof** and keep the wood from **rotting.**

Don't use it!

Shipbuilders do not use wood to build large ships such as oil tankers. They use metal instead. Metal is much stronger and can be made into huge pieces. Pieces of metal can also be joined more strongly than pieces of wood.

Preserving Wood

When wood stays wet over time, it **rots** away. It gets soft and falls apart. Rotting is a **natural** process. It happens because tiny **organisms** grow in the wood. They break apart the wood's **fibers** and feed on them. If we use bare wood outdoors it rots away quickly. Wood in the ground, such as fence posts, rots very quickly because it is always wet. Insects such as termites also eat wood, making it weak.

This rotted wooden roof will have to be rebuilt.

Treating wood

We make sure that wood does not rot by using **chemicals** called **preservatives.** We can paint preservatives on or dip whole pieces of wood in a bath of preservatives. Organisms cannot grow on the wood because the preservative is poisonous to them. We can also stop rot by **varnishing** the surface of wood.

Woodworm

Old floorboards and pieces of furniture often have tiny holes in them. This is called woodworm. The holes are bored in the wood by young beetles called larvae. They eat the wood as they bore through it.

The preservative on the brush will soak into the wood, protecting it from rot.

Weaving Wood

Small branches and thin strips of wood are **flexible.** They bend without snapping. This property makes them good for **weaving.** Weaving is done by passing the branches or strips of wood over and under each other. Panels for fences are made by weaving. Some panels are woven from wide, thin strips of wood. Others are woven from tree branches with their twigs and leaves stripped off.

This man is weaving a fence in a garden from bendable branches.

24

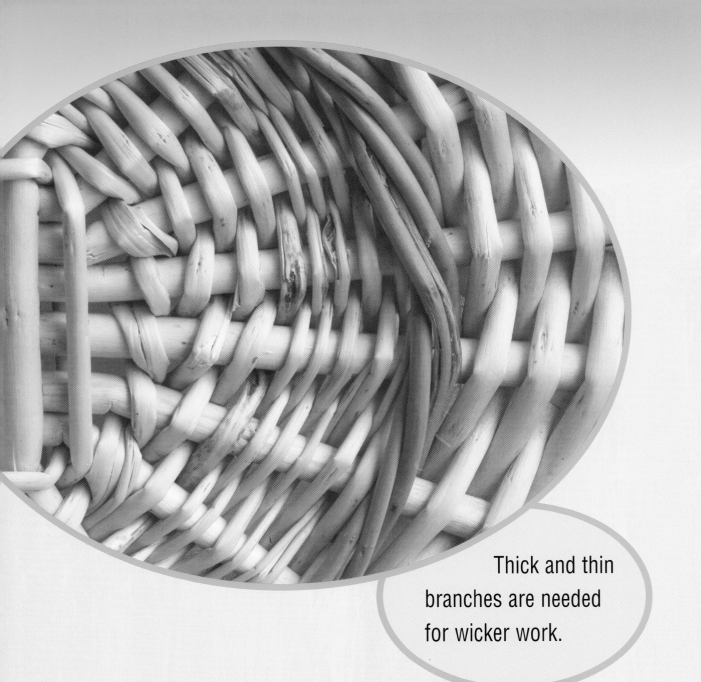

Thick and thin branches are needed for wicker work.

Wicker

We also weave wood to make chairs and baskets. Wicker is made by weaving thin branches from willow trees. Chairs made from wicker are light and strong. They also bend slightly, which makes them comfortable to sit on.

Wood for Objects

Wood is a good material for making all sorts of everyday objects. You can find wooden objects such as mixing spoons in the kitchen. Your home might be decorated with pictures in wooden frames. All these things could be made with plastic instead, but people often prefer wood because it looks **natural.** Some wooden things, such as lollipop sticks, are very cheap to make so we can throw them away after using them.

Bottle corks are made from the thick bark of cork trees.

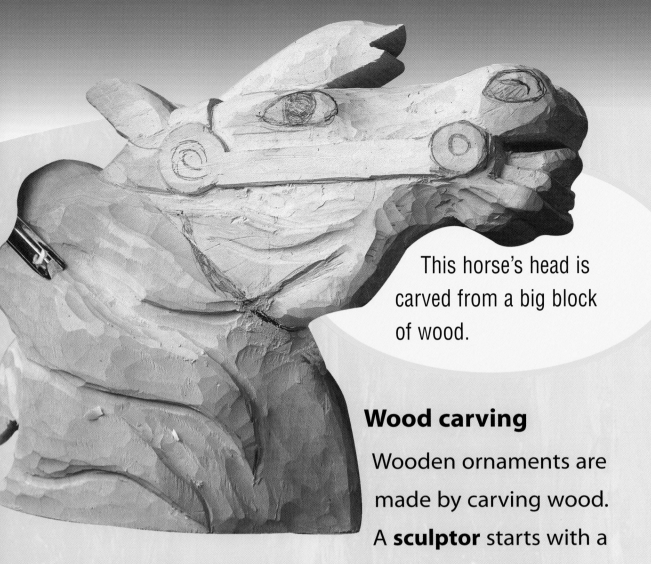

This horse's head is carved from a big block of wood.

Wood carving

Wooden ornaments are made by carving wood. A **sculptor** starts with a block of wood. He or she cuts away pieces of the block with a chisel and hammer, leaving behind the wood that makes the shape. Wooden furniture is often decorated by carving, too.

Don't use it!

We do not make very tiny or complicated shapes from wood. They are too difficult to carve. Tiny and complicated parts for machines are made from plastic or metal instead. Wooden parts would wear out too quickly.

Wood and the Environment

Wood is a **natural** material. We should never run out of wood because we can always grow more. However, some types of tree are running out because people have cut them down and not planted new ones. Many trees in the world's **rain forests** are being cut down for their valuable wood, like teak and mahogany. Some of these trees have taken more than one hundred years to grow and we can never replace them. Animals that live in the forest are also losing their homes.

Put newspapers and magazines into recycling bins.

These trees were specially grown for their wood.

Making paper from wood uses up a lot of energy. It also uses **chemicals** that are dangerous to throw away. We can save energy and chemicals by using paper again. This is called **recycling.**

Don't use it!
*We should not use wood without knowing where it comes from. We should only use wood that comes from forests that are specially grown for their wood. These are called **sustainable** forests.*

Find Out for Yourself

The best way to find out more about wood is to investigate for yourself. Look around your home for wood, and keep an eye out for where wood is used through your day. Think about why wood was used for each job. What **properties** make it suitable? You will find the answers to many of your questions in this book. You can also look in other books and on the Internet.

Books to read

Ballard, Carol. *Science Answers: Grouping Materials: From Gold to Wool.* Chicago: Heinemann Library, 2003.

Hunter, Rebecca. *Discovering Science: Matter*. Chicago: Raintree, 2001.

Using the Internet

Try searching the Internet to find out about things having to do with wood. Websites can change, so if one of the links below no longer works, don't worry. Use a search engine, such as www.yahooligans.com or www.internet4kids.com. For example, you could try searching using the keywords "forestry," "boat building," and "sustainable forest."

Websites

www.nationalgeographic.com/geographyaction/backyard/
A great site about a conservation program that focuses on United States public lands.

www.forestry.gov.uk
A fun site that explains how wood is grown and used.

Glossary

chemical substance that we use to make other substances, or for jobs such as cleaning

fiber long, thin, bendable piece of material

flexible describes a material that can bend without snapping

fuel material that burns well, giving out lots of heat. Wood is a good fuel.

grain pattern of lines in wood. Grain is made by the wood's fibers.

industry group of business organizations that do the same job. For example, forestry is an industry.

natural describes anything that is not made by people

organism living thing

preservative chemical that helps to stop wood from rotting

property quality of a material that tells us what it is like. Hard, soft, flexible, and strong are all properties.

rain forest thick forest that grows where the weather is always wet and warm

recycle to use material from old objects to make new objects

rot process that makes wood weak and crumbly. It happens when tiny organisms eat away the wood.

sculptor person who designs and makes figures out of wood or other materials

sustainable source of material that never runs out. Sustainable forests are grown to supply wood.

varnish liquid that is painted on wood to protect the wood

veneer very thin sheet of wood

waterproof describes a material that does not let water pass through it

weave to make fabric by threading fibers or yarns under and over each other

Index

DATE DUE

GAYLORD	PRINTED IN U.S.A.